Frogs and Toads

NATURE'S PREDATORS

Jeffrey O'Hare

KIDHAVEN PRESS™

San Diego • Detroit • New York • San Francisco • Cleveland
New Haven, Conn. • Waterville, Maine • London • Munich

© 2003 by KidHaven Press. KidHaven Press is an imprint of The Gale Group, Inc., a division of Thomson Learning, Inc.

KidHaven™ and Thomson Learning™ are trademarks used herein under license.

For more information, contact
KidHaven Press
27500 Drake Rd.
Farmington Hills, MI 48331-3535
Or you can visit our Internet site at http://www.gale.com

ALL RIGHTS RESERVED.
No part of this work covered by the copyright hereon may be reproduced or used in any form or by any means—graphic, electronic, or mechanical, including photocopying, recording, taping, Web distribution or information storage retrieval systems—without the written permission of the publisher.

LIBRARY OF CONGRESS CATALOGING-IN-PUBLICATION DATA

O'Hare, Jeffrey
 Frogs and toads / by Jeffrey O'Hare.
    p. cm. — (Nature's predators)
 Summary: Discusses the physical characteristics, habitats, behavior, and predators of frogs and toads, focusing on how they hunt and kill their prey.
 Includes bibliographical references (p.   )
    ISBN 0-7377-1388-7 (hardback : alk. paper)
 1. Anura—Juvenile literature. [1. Frogs. 2. Toads.] I. Title. II. Series.
 QL668.E2 O36 2003
 597.8—dc21   2002015471

Printed in the United States of America

Contents

Chapter 1:   What Is a Frog?   4

Chapter 2:   On the Hunt   18

Chapter 3:   Problems with Predators   32

Glossary   42

For Further Exploration   44

Index   45

Picture Credits   47

About the Author   48

Chapter 1

What Is a Frog?

Frogs and toads are amphibians. "Amphibian" comes from two Greek words meaning "a double life." Amphibians are born in water and spend most of their formative time in water. As they grow older, amphibians become land creatures, spending only a small amount of time in water.

Frogs and toads are so alike that most scientists do not make distinctions between them. However, there are ways to tell the difference between the two. Frogs usually have smoother, moister skin. True frogs are those that live both in and out of water. Aquatic frogs are slightly different; they are the ones who spend most, if not all, of their lives in water. Aquatic frogs have webbed toes. Frogs tend to lay their eggs in clusters. Frogs are usually leapers, covering good distances with powerful legs.

Toads usually have bumpier, drier, rougher skin. They live mainly on land. Toads do not have

Although frogs (top) and toads (bottom) look similar they have many differences.

webbing between their toes. They lay their eggs in long chains. Toads are hoppers, going only short distances on stubby, heavy legs.

Frog Facts

Like dinosaurs, frogs are prehistoric. Scientists believe frogs began to evolve from water animals almost 200 million years ago. Today, there are about three thousand known species of frogs and toads.

Frogs live on all the world's continents except Antarctica. Frogs and toads can live in a variety of environments, from high mountains to the trees of a rain forest or the floor of the desert. Frogs are even found above the Arctic circle. They can exist in almost any **habitat** as long as fresh water is nearby.

The world's smallest frog is the Cuban pygmy frog which measures about one-half inch long when fully grown. The Goliath frog of West Central Africa is the largest. It can grow to more than a foot in length. The largest species of toad is South Africa's Blomberg's toad, which can be as large as a dinner plate. A common adult bullfrog's body measures from six to eight inches.

True frogs are found on the ground. They make their homes in moist dirt, in small piles of leaves, or anywhere close to a source of moisture. Aquatic frogs have webbed feet that help propel them when swimming. Most species of tree frogs have no webbing between their toes. Tree frogs have round sticky pads at the end of each toe that help grip wet leaves and slick branches.

Life Cycle of Frogs and Toads

1. Eggs are laid.

2. A tadpole breathes using gills.

3. A tadpole develops legs.

5. The adult frog or toad loses its tail.

4. The almost mature frog or toad (at 2-4 months) still has some of its tail and breathes using lungs.

The flying frogs of Borneo and Costa Rica are tree dwellers that have webbing under the arms as well as between the toes and fingers. This webbing unfolds like small sails when the frogs leap from a branch. These frogs glide on air currents, moving from one tree to another to escape **predators** or to look for food.

Hot and Cold

Like all amphibians, frogs and toads are **ectothermic**, or cold-blooded, animals. This means they cannot keep their body temperature stable. Their body temperature reacts and changes depending on the outside temperature.

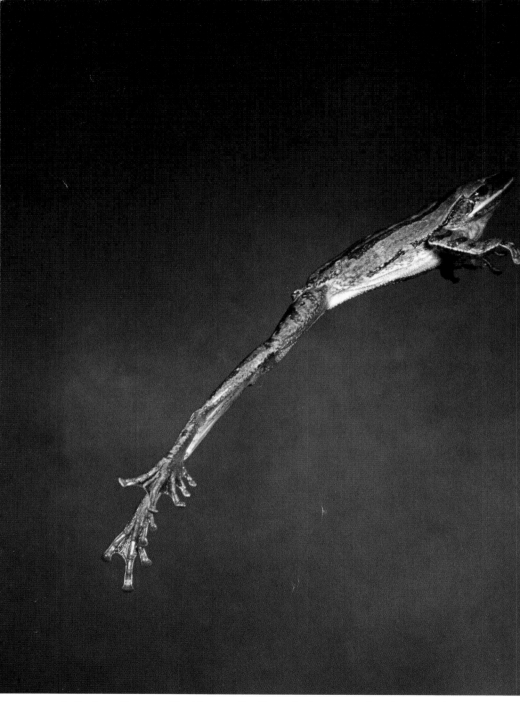
A flying frog sails through the air in search of a meal.

Frogs **hibernate** in cold weather. Some dig burrows in the ground or in the mud at the bottom of ponds to prepare for a long sleep. They hibernate until spring, when the temperature grows warmer. When it gets very cold, a frog's activity decreases. The North American wood frog survives for months in a frozen, deathlike condition. This frog turns the **glucose**, or sugar, in its body to a kind of antifreeze that concentrates in the frog's vital organs, saving them from damage while the rest of the frog freezes. Frogs have survived temperatures as low as 21 degrees Fahrenheit.

Desert frogs and toads go through a process called **estivation**, which is similar to hibernation. When the temperature becomes too hot, they burrow in the ground to sleep until things cool off.

In either case, the resting frog does not need to hunt or eat. Its body slows down or stops completely so the frog does not have to eat. When the ground around the frog returns to more normal temperatures, the frog wakes up. The frog then digs itself out and immediately starts looking for a meal.

Meat Eaters

Most of the world's frogs, including all North American frogs, are **carnivores**. Very few frogs are **herbivores**. Izecksohn's Brazilian tree frog in South America eats fruit. Another rare exception, the *Bufo marinus* toad, eats meat and some vegetation.

A hungry frog gobbles up a dragonfly.

Carnivorous frogs eat small animals, such as invertebrates (which have no spine) like ants, termites, beetles, slugs, worms, and snails. They also eat flies, mosquitoes, small crustaceans, spiders, centipedes, and fish. Sometimes they eat slightly larger animals such as ducklings, snakes, baby turtles, bats, and mice. Larger frogs may also be **cannibals** and eat other frogs.

All **prey** is swallowed live and whole, then slowly digested within the frog's stomach. Since frogs swallow their prey whole, they can eat almost anything that will fit in their mouths.

Get a Move On

Almost all species of frogs and toads are attracted to their prey through movement. This means the prey must be alive. Some toads will eat hamburger or other "dead" material, but only as a small part of their overall diet. **Tadpoles** and some toads will scavenge off a dead animal if that is the only food available. Given a choice though, a toad always prefers live food.

When frogs detect prey, they slowly rise to get ready to attack. The Argentinian horned frog for example, twitches its front toes to entice the prey to come closer. When the prey is within range, any movement triggers a response in the frog's brain to attack.

If the prey stops moving long enough, the frog may turn away. A fly that can stay still will avoid being eaten. But the slightest movement of a leg or wing will cause the frog to strike before the prey can escape.

Any Drinks with the Meal?

Although a frog eats a wide variety of prey, it never drinks. Moisture is absorbed through its skin by **osmosis**. Osmosis is one reason frogs stay near water. It is also why they shed their skin. When the skin becomes blocked with dirt or other substances, moisture or oxygen cannot pass through it. Without clean skin, a frog will die.

Some frogs can make do with very little water. The water-holding frog of Australia's deserts can live up to seven years without rain or new water. Inside its burrow, the frog surrounds itself in a **sac** of skin to seal in what little moisture is available.

Friend or Foe?

Frogs and toads benefit the environment. The majority of a toad's diet consists of insects and other pests. Gypsy moths, army worms, tent caterpillars, slugs, and other destructive creatures are all delicious to a toad. By eating insects and rodents, frogs keep the population of these pests low. However, not all toads are welcome guests.

When frogs or toads are introduced to a new environment, they may become a problem. When Australia imported the cane toad to help with their beetle population in the 1930s, it was a disaster. The toads were **nocturnal**, coming out only at night, while the beetles were active during the day. Since there were no beetles to eat at night, the toads quickly ate all sorts of other wildlife, including birds and snakes. There were no natural predators to keep the toads in check. They grew to an overwhelming number and still remain a problem today.

Even in their natural surroundings, a troop of frogs can cause problems. Preferred prey animals may be quickly reduced, or wiped out altogether, by large numbers of frogs all eating the same thing.

A tiny gray mouse peers hopelessly out of a frog's mouth.

A Troop Together

In the proper environment, where the natural checks and balances of the **ecosystem** are present, a troop of frogs can thrive nicely. As long as there is plenty of food to eat, and enough natural predators keep the frog population from growing too large, life goes on without major problems.

Frogs do not normally travel in troops. But when ponds dry up, they search for food more actively. They may move together from a former feeding ground to a new area, or from a dry area to water.

When they are on the move, frogs become easier targets for predators. While struggling to find more food, they may become a meal for something else.

In the Beginning

From the early stages of life, a frog is an eating dynamo. A small, squiggly, bloblike creature, a tadpole is built to feed.

When a tadpole hatches, it has many tiny **denticles**, or small teeth, around its mouth. These are used for scraping algae and other food from bottom debris or directly from the water. A tadpole's strong mouth is designed to hang on to rocks, sticks, plants, or other objects in the water while it scrapes off food particles. A tadpole can hang upside down in the water and wave its tail back and forth to create a flow of water that funnels algae and bacteria into it's mouth.

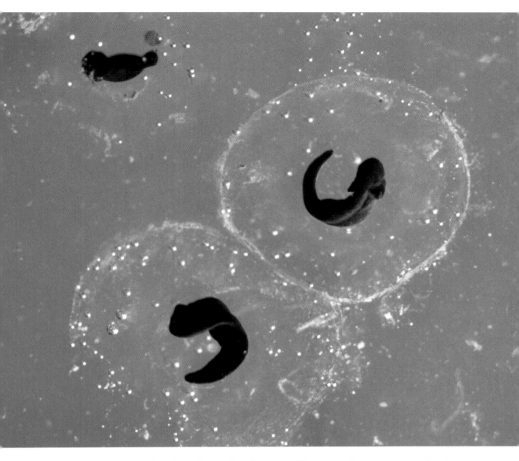

Once these tadpoles hatch they will spend most of their time feeding on a variety of algae and bacteria.

Tadpoles are filter feeders. They gulp water which passes through the mouth to be filtered out through specialized organs. Food is strained and kept in the body for nourishment, while the water is expelled through the gills.

Tadpoles will eat almost anything, including plants, decaying animals, or other tadpoles. They

A bullfrog gulps down a carpenter frog. Frogs will eat just about any animal they can swallow, including other frogs.

also eat a lot, almost constantly skimming the water for food. Some frogs lay unfertilized eggs among the tadpoles so the young can feed on the eggs. The female harlequin poison dart frog, for example, visits her spawn every few days to drop such eggs for food. Tree frog tadpoles have heavily muscled mouths that allow them to suck in the whole eggs of other species.

The two main visible features of a tadpole's body are the tail and the mouth. Movement and eating are a tadpole's only needs. As frogs and toads mature, they become less frantic in their active search for food. Frogs can travel great distances to find food, but they prefer that food comes to them. Frogs' hunting strategies, then, involve a lot of hiding and waiting for prey. Even so, frogs and toads are efficient predators.

Chapter 2

On the Hunt

Frogs and toads hunt in many different ways. Some frogs hunt during the day, while others, like the crawfish frog, are nocturnal. The cricket frog gets by on an average of thirteen insects a night. In comparison, an American toad will fill and empty its stomach four times in each twenty-four hour period.

Most frogs use a quick tongue to snare prey, while some frogs have no tongues. Frogs that do not use their tongues must rely on natural camouflage and the ability to remain still until prey approaches. Then they use strong legs to pounce on the victim.

Even with all these variations, frogs and toads have one thing in common. They eat continuously, hunting for food the entire time they are awake. However, frogs cannot overeat at any particular moment. Because frogs take in their prey whole,

A gray tree frog snatches a cricket and swallows it whole.

one item must be forced into the throat before the frog can move on to a second course.

Frogs are solo hunters. They feed only themselves. Except for females who leave eggs for tadpoles, parents do not feed their young. Different breeds of frogs can exist together, hunting next to one another.

Frogs do not normally share food because prey is usually swallowed right away. However, when two frogs of different sizes come into conflict, a bigger frog may simply swallow a smaller frog, captured prey and all.

The Easy Way

The main hunting tactic of most frogs is ambush. Frogs sit and wait quietly for prey to come near. Nocturnal frogs wait near a light source on warm summer nights to lap up bugs that are attracted to the light.

When hunting, the frog sits virtually motionless for long periods of time. It slows its **respiration** down until its sides do not move. To insects, it seems like any other lump in the underbrush. That is when the frog strikes, to grab the unsuspecting meal.

Frogs who eat larger prey are more cunning. A mouse or duckling is equal in size and speed to a frog. When going after such prey, the frog is more patient. Frogs and toads without tongues open their mouths as wide as possible and spring out to swallow prey.

Smaller frogs are quicker and do not remain hidden during the hunting process. Many smaller frogs launch themselves at their victims. This assault happens very swiftly. This makes for quick strikes and quick swallowing of small bugs and animals that might otherwise be fast enough to move away.

Natural Camouflage

Frogs have developed various colorings to help them hide from predators that eat frogs. This camouflage also allows the frog to lie in wait for unsuspecting prey.

In its natural habitat, the Asian horned toad waits half-buried in leaves. Its dark, mottled skin coloring makes the toad almost impossible to see.

A leaflitter toad blends in with a leaf in the Amazon rain forest.

The leaves explode with life when the toad jumps out to swallow its catch.

Leaf frogs of the Solomon Islands not only have the natural coloration of the leaves, their bodies also have adapted to the pointed shape of the local **flora**. They look exactly like any of the leaves scattered on the ground.

Some frogs can change their coloring in response to outside stimuli like temperature or an attack. Other frogs change their coloring to blend in with their surroundings. The African arum frog, for example, has ivory-colored skin. This coloring allows the frog to live within the white blossoms of the arum swamp lily where it laps up insects who come to sniff at the flower. When the arum lily is not in bloom, the frog's skin coloring changes, allowing it to hide among other flowers.

Waiting to Attack

The most colorful frogs are found in the world's tropical rain forests. These frogs range from hot orange to cool blue. Many tropical tree frogs hide among brightly colored fruits, waiting for insects to come by. Then the frogs attack in the same manner as their more northern relatives.

Not all frogs are colorful or take advantage of the ambush. The crucifix frog, for example, likes to eat termites and ants. It waits beside an ant trail for long periods of time, licking up ants as they march by.

The European common toad is another hunter who works in plain sight. It sits by the entrance to a

A bright orange frog sits motionless as it waits to attack an unsuspecting insect.

beehive at the end of the day, eating returning bees. Its strong, warty skin is too thick to be bothered by the stings and **toxins** of the outraged worker bees.

Body Work

Almost every part of a frog's body is designed to make it an able predator. Frogs with teeth do not use them to chew food. The purpose of these teeth is to hold captured prey in place so it cannot wriggle out of the frog's mouth.

Some species of frogs have no teeth, while others have teeth that grow quite large. The big horned frog of South America is one of the only frogs whose teeth get large enough to inflict a nasty bite when it clamps onto its prey.

The tusked frog of Australia is an unusual breed with a set of false tusks on its lower jaw. These tusks are used for defensive purposes, not for hunting or eating.

But the teeth are only a supporting part of the mouth. The real action is in the tongue.

Tip of the Tongue

A frog's tongue is a deadly weapon. In less time than it takes to blink, a frog unrolls its tongue, shoots it out to snag nearby prey, then rolls it back inside its mouth. This movement is so fast, it can hardly be seen by the human eye. Scientists have been able to study the actions of a frog's tongue only through the use of special cameras and high-speed photography.

The frog's tongue is attached to the front of its mouth and folds back inside. When a frog attacks, the tongue comes forward like a human arm does when flipping a yo-yo. The part of the tongue that strikes the prey is similar to the back of the hand, not the palm.

A South American giant toad prepares to unroll its tongue to catch a worm.

Once the prey has been hit, the frog folds its tongue back into its mouth. It is the tongue's **pliability** that makes it seem to roll back in like a party favor.

A few species of frogs do not use their tongues to capture food. Instead of a sticky tongue, they have strong arms and legs. Aquatic frogs and clawed frogs have no tongues. Such frogs sit motionless in the water, with their arms and hands stretched wide, waiting to feel motion in the water.

Fire-bellied toads have short tongues that cannot be extended beyond their mouths. The toad uses its hands to scoop the prey toward its mouth, while lunging forward so the prey can be pushed inside the body.

The African clawed frog has no tongue. It stuffs a meal in its mouth, then clamps its jaws closed to keep any prey from slipping loose. This can be difficult since these frogs prefer larger prey, such as mice or other frogs.

The Eyes Have It

Being able to spot darting insects requires good eyesight. A frog's big eyes are perfect for hunting. Frogs can see a wide variety of colors in sunlight, and they see well in dim light.

Since the eyes are located high on their heads, frogs are able to sit in the water with only their noses and eyes exposed to the air. This allows the frog to remain hidden in the water while waiting for insects to pass by.

A frog watches for prey, its eyes high atop its head and able to see in all directions.

Frogs also can see in almost a complete circle. It is difficult to sneak up on a frog because they can see predators coming from any direction.

Frogs' eyes have a membrane that acts as a shield to protect the eye from wind, water, or dirt, while keeping the eye moist. The membrane is important to the hunting frog because it acts as a kind of camouflage. A frog can see through the membrane, but prey cannot tell where the frog is looking.

The Surinam toad has small eyes and hunts by touch. It sits in murky water on the bottom of rivers and ponds, with its forefeet spread out before it. When a fish or insect passes over its sensitive fingers, the frog lunges forward to shove the prey into its mouth.

On the Inside

Once prey is caught, a frog's jaws clamp together tightly and its teeth hold the animal in place. A frog may use its hands to reposition captured prey in order to make it easier to swallow.

A larger animal that is grabbed by the hind end may live a long time while being eaten and digested. If an animal is particularly large, the frog might not be able to digest it all at once.

Prey that goes in head first quickly drowns in the frog's salivary juices or suffocates inside the frog's body. Digestive enzymes break the food down into digestible form. Every piece of the swallowed prey passes through the frog's digestive system. Usable material breaks down in the stomach. Minerals, fats, and other vital juices are absorbed into the frog's body. The remainder of the swallowed prey, including the bones, moves through the intestines to be passed out of the body as waste. If the bones prove too difficult to digest, they may be eliminated from the body during a process known as stomach cleaning.

A Belly Ache

The feeding impulse is so strong in frogs that they sometimes swallow things without waiting

A cricket suffocates under the pressure of a toad's clamped jaw.

to determine how they taste. When a frog eats something poisonous or bad tasting, it can vomit its own stomach. With a mighty heave, its body turns almost inside out, pushing the stomach lining outside the frog's mouth.

Dropping their eyes into a cavity above the roof of the mouth helps frogs move food into their stomachs.

Once the stomach is out, the frog cleans it by using its right front leg to rub off the offending substance. When the stomach lining is clean, the frog pushes the stomach back inside its mouth. A few heavy swallows gets the stomach back in place.

The Blink of an Eye

Just as they are important aids during the hunt, a frog's eyes serve an important purpose in the eating process. When a frog needs to swallow its prey, the eyes move down from their usual position. The muscles that hold the eyes in place relax, letting the eyes sink inward through special openings in the skull. The eyeballs sink into the head to a cavity above the roof of the mouth. Once squeezed inside, the eyeballs help push the prey further into the throat. This is why frogs blink when they are eating.

Each part of a frog or toad is designed to make it an efficient predator. But frogs and toads are not immune to being hunted and eaten by other predators.

Chapter 3

Problems with Predators

Frogs and toads face a wide variety of dangers, including predators. Vast numbers of frogs' eggs never make it to be tadpoles. Weather, temperature, water flow, and pollution can destroy the eggs. There are also plenty of predators to eat the eggs, including other tadpoles. Dragonfly larvae and other insects, as well as a wide variety of fish, will feast on frogs' eggs.

A swimming tadpole attracts predators such as turtles and snakes, as well as other tadpoles and frogs. As a frog matures, it grows into a bigger target for bigger predators.

Birds and snakes are especially dangerous to frogs and toads. The crested hawk, the egret, and the heron are birds that feed primarily on frogs. Rat snakes eat frogs, and the hognosed snake feeds on toads. Other predators include raccoons, otters, bass, alligators, crocodiles, bats, snapping turtles, and other frogs.

A Good Defense

Frogs and toads are hunted in many ways. Birds hunt by sight while snakes hunt by smell or by detecting body heat. Since a frog faces so many dangers, nature has provided a number of defenses the frog can use to escape predators.

When a predator is near, frogs and toads remain still, hoping they will not be seen or smelled. A frog can even stop breathing, holding its breath

A slithering snake dines on a cluster of frog's eggs.

until a predator moves on. If a frog is detected, it jumps to get away.

Frogs are powerful jumpers. Bullfrogs can leap up to six feet. A leopard frog, which may be only half the size of a bullfrog, can cover almost the same distance. True frogs, who live close to water, tend to stay within a few leaping distances of the water. When they need to escape, they jump into the water and swim away.

Other Defenses

Frogs and toads have other defenses to use when cornered. Many frogs gulp in air to make themselves look bigger and more imposing. The tomato frog inflates to get big and fat to convince predators they will be unable to swallow it.

Some other species of frogs become aggressive when facing a predator. They leap directly at the

---

A tiny green tree frog sits motionless in the jaws of a hungry crocodile.

predator. They attack with their mouths open, even if they have no teeth with which to bite. Many frogs scream while attacking, being attacked, or leaping away. This scream is disorienting enough to confuse a predator.

Still other frogs play dead to get predators to leave them alone. The leopard frog holds its breath and lies flat. The predator thinks the frog is dead and drops it on the ground.

A frog can also release its bladder while jumping. The bladder is an internal sac where urine is stored. When the bladder is released, the urine spray covers the frog's scent as the frog leaps away. The urine tastes awful to any predators.

On the Markings

Most frogs have distinctive coloring they "flash" to confuse predators. While a frog, such as the pickerel frog, may look plain from the back, it has bright colors or markings on the belly or under the legs. When the frog leaps away, these bright colors become visible. The colors give the predator something to focus on. When the frog lands, its legs fold under and the colors are hidden. The predator momentarily loses sight of the frog because the colors have disappeared. This gives the frog an extra few moments either to sit still or make a second leap away.

Other helpful colorings that assist frogs in avoiding predators include special body markings. A few South American frogs, like the Cuyaba dwarf frog, have large spots on their backs. When the frog inflates itself, the markings look like the eyes of a

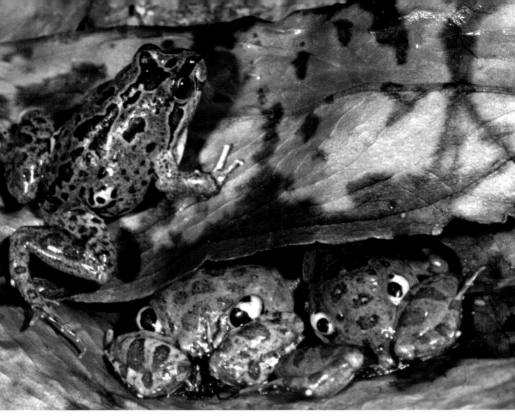

The spots on these false-eyed frogs confuse predators by making them think they are the eyes of a large animal.

larger animal. This is usually enough to scare predators away.

The fire-bellied toad is normally gray. When frightened or alarmed, the toad bends backward, raising its hands and feet in a weird bowl-like contortion. With this movement the frog flashes bright orange spots on its stomach, which warn predators to stay away.

One breed of South American tree frog is black, white, and gray. When the frog remains still, it resembles a pile of bird droppings. Since no animal eats bird droppings, the frog is usually left in peace.

Pretty Poisons

Some frogs, especially those found in tropical areas of South America, produce poisons. Bright orange, yellow, red, and blue skin coloring convey a warning to predators that these frogs are better left alone. South America's small, colorful frogs are also some of the world's deadliest. Any animal unlucky enough to eat one of these frogs will quickly die. The toxins of a poison dart frog are so powerful that simply picking one up with a bare hand will cause the skin to tingle, and even a small amount of the poison is enough to kill a human.

These colorful frogs and toads are not the only species that produce poisons. The Colorado River toad is a dull gray or green color. It has warty glands on its legs and near its mouth that ooze poison that is toxic to both animals and humans.

Many toads have glands that secrete a poisonous white substance when they are attacked. The common toad's poison can cause a dog to foam at the mouth and vomit. The powerful secretions of the giant toad have been known to kill dogs and people who have bitten or eaten this toad. Some frogs have several different poisons on their bodies, including one for stunning prey and one for warding off predators.

None of these frogs or toads bite to inject their poisons, and only a few toads can squirt poison at an attacker. They pass these poisons to attackers through direct contact with the skin.

The powerful poisons in a frog's body are used solely for defensive purposes. For the frog, these poisons have to be triggered by some outside assault. Frogs and toads do not use their poisons for hunting.

Any poisons must be squeezed or sweated out of a frog in order for humans to make use of them. South American natives know how to keep a frog over a fire so the poison is sweated out of the body. This poison is so toxic that it remains deadly even after having been exposed to air.

While frogs and toads have poisons to protect themselves, there are predators who have figured out ways around the toxins. The gopher frog eats toads, but immediately ejects any poisons from its mouth. If the frog is in the water when this happens, large foamy masses of poison float to the sur-

---

The bright blue color of this poison arrow frog warns predators to stay away.

face. Skunks eat toads, but they roll the toads roughly across dirt or grass until all the poison is released from the body. Some breeds of snakes also seem unaffected by these poisons. Even though all these animals can eat frogs or toads, none of them is the frog's most deadly enemy.

Frogs and Humans

Of all the dangers that frogs and toads must face, none is deadlier than humans. People harm frogs in a variety of ways.

Humans hunt frogs as a food source. Frog legs are considered a delicacy in some parts of the world. Bullfrogs are popular food sources in areas of the United States. The islands of the Caribbean have a large frog they call the "Mountain Chicken," because it is so tasty.

Besides being eaten, numerous frogs are bred or captured for classroom experimentation. Many junior high and high school science classes dissect large numbers of frogs each year, even though many computer programs and other teaching tools exist that can better explore frog anatomy.

People who keep exotic pets are another danger to frogs and toads. Frogs are colorful and easy to care for in the right environments, so they make popular pets. Many countries have outlawed the export of native animals, including frogs, as pets or for private collectors. But as long as there are buyers for these animals, dealers will continue to capture frogs and remove them from their natural habitats.

These habitats are destroyed when people move into and develop new areas. Tropical rain forests, important habitats for a large variety of frog species, are particularly in danger. But they are not the only places that are in trouble. In the

---

Frogs are killed and served as delicacies in many places around the world.

Five red-eyed tree frogs rest on a branch in a rain forest in Central America.

United States, wetlands are disappearing at an alarming rate because old rivers are dammed up or rerouted and swamp lands are drained for commercial development.

Pollution from acid rain and pesticides also has a devastating effect on amphibian populations around the world. Frogs and toads give indications about the overall health of a particular environment. As changes appear in the frog population, scientists try to figure out what is causing these variations. The scientists study what effect the causes of such changes will have on humans.

Healthy frogs indicate a healthy environment. Wetlands and rain forests, which are the breeding grounds and living areas for many frogs and toads, are important to the health of the entire world.

Glossary

cannibal: An organism that eats members of its own species.

carnivores: Meat-eating organisms.

denticles: Microscopic toothlike projections.

ecosystem: A community of organisms and its environment functioning as an ecological unit.

ectothermic: Cold-blooded.

estivation: Period of rest during extremely hot weather.

flora: The natural vegetation of an environment.

glucose: Natural sugar produced by an organism.

habitat: The particular environment in which an organism lives.

herbivore: An organism that eats only plants.

hibernate: To rest, for a period of time, during extremely cold weather.

nocturnal: Active primarily at night.

osmosis: The process that allows one substance to pass through a barrier.

pliability: The ability to bend freely or repeatedly without breaking.

predator: An animal that hunts other living organisms.

prey: An animal hunted for food.

respiration: The use of oxygen by an organism's cells and tissue.

sac: Part of the body that acts like a pouch or bag.

tadpole: A frog or toad larva that has a rounded body and long tail.

toxin: A poisonous substance.

For Further Exploration

Densey Clyne, *It's a Frog's Life*. Milwaukee, WI: Gareth Stevens, 1998. Part of the Nature Close-Ups series. Short book at an informative, easy-to-read level.

Mary C. Dickerson, *The Frog Book*. New York: Doubleday, Page, 1906. A very old book that offers a complete examination of frogs and toads. Features numerous photographs and illustrations by the author.

Mike Linley, *The Frog and the Toad*. Ada, OK: Garrett Educational Corporation, 1992. One volume of the Wildlife Survival Library, this is an easy to read book that explores the complete life cycle of the frog. It features big pictures of a wide variety of frogs.

Chris Mattison, *Frogs and Toads of the World*. New York: Facts On File, 1987. A very thorough and scientific book on frogs and toads.

Edward R. Ricciuti, *Amphibians*. Woodbridge, CT: Blackbirch, 1993. One of Our Living World series, this volume offers a quick but informative read, with diagrams and beautiful photos.

Index

African arum frogs, 22
aggression, 34–35
American toads, 18
amphibians, 5
aquatic frogs, 6, 26
Argentinian horned frogs, 11
Asian horned toads, 21–22

bees, 24
Blomberg's toads, 6
bodies
  differences between frogs and toads, 4
  shape of, 22
  temperature of, 7
Bufo marinus toads, 9
bullfrogs
  as food for humans, 39
  movement of, 34
  size of, 6

camouflage
  as defense, 21–22
  eye membrane and, 27–28
  hunting and, 18
cane toads, 12–13
cannibals, 20
carnivores, 9, 11
clawed frogs, 26
Colorado River toads, 36
coloration
  defense and, 35–36
  hunting and, 22
  poisons and, 36
crawfish frogs, 18
cricket frogs, 18
crucifix frogs, 22
Cuban pygmy frogs, 6
Cuyaba dwarf frogs, 35–36

defenses
  aggression as, 34–35
  camouflage as, 21–22
  coloration as, 35–36
  eyes as, 27–28
  movement as, 34, 35
  playing dead as, 35
  poisons as, 36–39
  respiration as, 33–34
  tusks as, 24
  urine spray as, 35
denticles, 14
diet
  described, 9, 11, 12, 21, 26
  frequency of eating and, 9, 18
  of tadpoles, 14–15, 17
digestion, 11, 18, 20, 28–29
drinking, 12

eating, 9, 17, 31
  see also digestion
ectotherms, 7
eggs, 5, 17, 32
environment, 12–14, 40–41
estivation, 9
European common toads, 22, 24
evolution, 6
experiments, 39
eyes
  defense and, 27–28
  eating and, 31
  hunting and, 26–28

feeding impulse, 29–30
feet, 6–7
filter feeders, 15
fingers, 6–7
fire-bellied toads, 26, 36
flying frogs, 6–7

45

food
  described, 9, 11, 12, 21, 26
  frequency of eating and, 9, 18
  of tadpoles, 14–15, 17
frogs, 5
  see also names of specific frogs

giant toads, 37
Goliath frogs, 6
gopher frogs, 38–39

habitats, 5, 6, 40–41
harlequin poison dart frogs, 17
herbivores, 9
hibernation, 9
homes, 6
horned frogs, 24
humans
  environment and, 40–41
  poisons and, 36, 37
  as predators, 39
hunting
  camouflage and, 18
  coloration and, 22
  eyes and, 26–28
  nocturnal, 12–13, 18, 21
  solo, 20
  tactics, 11–12, 20–21, 22, 24
  tongues and, 18, 24–26

Izecksohn's Brazilian tree frogs, 9

leaf frogs, 22
legs
  described, 4
  hunting and, 18
  stomach cleansing and, 31
leopard frogs, 34, 35

moisture, 12, 14
"Mountain Chicken" frogs, 39
movement
  as defense, 34, 35
  of frogs, 4, 6–7
  of prey, 11–12, 26
  of tadpoles, 17
  as targets during, 14
  of toads, 4
  of tongue, 24–26

osmosis, 12

pesticides, 41
pets, 39
pickerel frogs, 35
poison dart frogs, 17, 36

poisons, 36–39
pollution, 41
predators
  of frogs and toads, 32–33, 39
  need for, 13
prey
  digestion of, 11
  frogs and toads as, 13, 14, 32–33
  movement of, 26
  types of, 9, 11, 12, 21, 26
pygmy frogs, 6

respiration, 20, 33–34

sight, sense of, 26–28
sizes, 6
skin
  as defense, 21–22, 35–36
  of frogs, 5
  poisons and, 36, 37
  shedding of, 12
  of toads, 5
South American tree frogs, 36
species, number of, 6
stomach cleansing, 29–31
Surinam toads, 28

tadpoles, 14–15, 17, 32
taste, sense of, 29–30
teeth, 24, 28
toads, 5
  see also names of specific toads
toes
  hunting tactics and, 11
  padded, 6
  webbed, 4, 6–7
tomato frogs, 34
tongues, 18, 24–26
touch, sense of, 28
tree frogs
  eating by tadpoles of, 17
  Izecksohn's Brazilian, 9
  South American, 36
  toes of, 6
tropical rain forests, 22, 40, 41
tusks, 24

urine spray, 35

vision, 26–28
vomiting, 30–31

water-holding frogs, 12
webbing, 4, 6–7
wetlands, 41
wood frogs, 9

Picture Credits

Cover Photo: © Carmela Leszczynski/AnimalsAnimals/Earth Scenes
© Jonathan Blair/CORBIS, 34
© Ron Boardman; Frank Lane Picture Agency/CORBIS, 15
© Tom Brakefield/CORBIS, 23
Corel Corporation, 16, 29
© Philip James Corwin/CORBIS, 27
© Michael & Patricia Fogden/CORBIS, 21, 33
© Michael Freeman/CORBIS, 40
© Jeffrey Lepore/Photo Researchers, 5 (top)
© Carmela Leszczynski/AnimalsAnimals/Earth Scenes, 19, 36, 41
© Joe McDonald/CORBIS, 38
© Tom McHugh/Photo Researchers, 13
© Gary Meszaros/Photo Researchers, 10
Chris Jouan, 7
© David A. Northcott/CORBIS, 8, 25
PhotoDisc, 30
© Kevin Schafer/CORBIS, 5 (bottom)

About the Author

Jeffrey O'Hare is the author of a number of non-fiction nature books. He has also written over one hundred different puzzle books for young readers. Other work includes writing for children's television shows and comic books. He lives in Bethany, Pennsylvania, with his wife and five children.